Education
252

天上的珍珠

Pearls in the Sky

Gunter Pauli

[比] 冈特·鲍利 著

[哥伦] 凯瑟琳娜·巴赫 绘

何家振 译

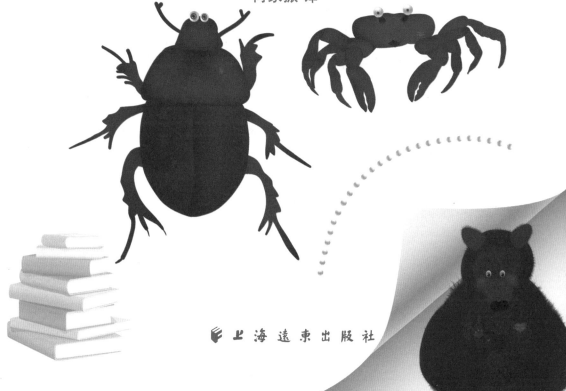

上海远东出版社

丛书编委会

主　任：贾　峰

副主任：何家振　闫世东　郑立明

委　员：李原原　祝真旭　牛玲娟　梁雅丽　任泽林

　　　　王　岢　陈　卫　郑循如　吴建民　彭　勇

　　　　王梦雨　戴　虹　靳增江　孟　蝶　崔晓晓

特别感谢以下热心人士对童书工作的支持：

匡志强　方　芳　宋小华　解　东　厉　云　李　婧

刘　丹　熊彩虹　罗淑怡　旷　婉　杨　荣　刘学振

何圣霖　王必斗　潘林平　熊志强　廖清州　谭燕宁

王　征　白　纯　张林霞　寿颖慧　罗　佳　傅　俊

胡海朋　白永喆　韦小宏　李　杰　欧　亮

目录

Contents

圣诞岛红蟹开始了穿越印度洋的大规模迁徙。一只蜣螂观察着这段长途跋涉。

蜣螂说："这么说，你们等了整整4年才踏上这趟旅程，迁徙到海里某个恰到好处的地点产卵。"

The Christmas Island red crabs are setting off on their large migration across the Indian Ocean. A dung beetle is observing the trek.

"So, you waited a whole four years to take this trip, to migrate to the exact right spot in the sea to deliver your babies," the dung beetle comments.

红蟹开始迁徙……

Red crabs are setting off on their migration …

成千上万的红蟹在月亮的指引下……

Millions of crabs, guided by the Moon ...

"没错，至少有一亿只红蟹一起决定去哪里，每只蟹都要产成千上万的蟹卵。我们仰望天空，以确定正确的出发时间。"红蟹回答道。

"哇！成千上万的红蟹在月亮的指引下，产下成千上万的卵。令人震惊！我们一生一般也就产几十个卵。我们靠银河导航。"

"Yes, at least one hundred million of us decide together where to go, to each release our millions of eggs. We look at the sky to figure out the right time to set off," the red crab responds.

"Wow! Millions of crabs, guided by the Moon, releasing billions of eggs. Impressive! We mostly lay only a few dozen in our lifetime. We find our way guided by the Milky Way."

"是吗？听起来太棒了！所以，你们不是靠一颗星星或月亮寻路，而是靠银河系寻路。"

"是的，我们仔细地注视着天空中那一道由许许多多星星组成的朦胧的光带。"

"你注意到这些新星了吗，那些看起来像一串珍珠的星星?"红蟹问。"不知它们从哪儿冒出来的……"

"Do you? That sounds fantastic! So, you are not guided by one star, or the Moon, but you travel following directions given to you by this galaxy."

"Yes, we carefully watch that hazy band of light in the sky, one that is formed by millions of individual stars."

"And have you noticed these new stars, the ones that look like a string of pearls?" Red Crab asks. "They just emerged out of nowhere…"

靠银河系寻路……

Directions given to you by this galaxy ...

……那是一串卫星……

... that is a string of satellites ...

"那些不是星星，那是一串卫星。"

"卫星？为什么在地球上会看到它们？"

"你知道，在这个地球上，谁想干什么就干什么。就像宇宙的西部荒野。"

"但那些亮晶晶的卫星会造成混淆。'星星'在地平线上升起，但实际上根本就不是星星？"

"Those are not stars, that is a string of satellites."

"Satellites? How is it that they are visible from Earth?"

"Here on Earth everyone does just as he pleases, you know. It is like the Cosmic Wild West."

"But those super shiny satellites will create confusion. 'Stars' emerging on the horizon – that are not stars at all?"

"一些卫星工程师试图把那些人造卫星涂成深色来降低它们的亮度。"

"那不管用！"

"为什么不管用呢？"蜣螂问。"我就不认为它们是银河系里的一组新恒星，因此它们不会影响我们的导航。"

"Some satellite engineers try to dampen the brightness of their satellites by giving some of their space objects a dark look."

"That does not help!"

"Why not?" Dung Beetle asks. "I, for one, don't think that those are part of a new set of stars in the Milky Way, which may affect our navigation."

那不管用！

That does not help!

......对地球生命构成威胁。

... a threat to life on Earth.

"世界各地的人们都在观察天空，寻找可能对地球生命构成威胁的小行星和陨石。当一颗大家伙撞上地球时，我们可能都会死。"

"我不明白！这和深色的人造卫星有什么关系？"

"There are people all around the world watching the sky, looking for asteroids and meteorites that may pose a threat to life on Earth. When a big one smashes into the Earth, we could all die."

"I don't get it! What does this have to do with dark-looking satellites?"

"这样说吧，人们望着天空可能是寻找光线，也可能是寻找热量。望远镜有两种：一种用来寻找光，另一种用来寻找热！"

"救命啊，我晕了！这与那些深色的人造卫星有什么关系？"

"东西暗了会怎样？"红蟹问。

"Well, you can look at the sky either looking for light, or looking for heat. There are two types of telescopes: one looking for light, the other for heat!"

"Help me, I am lost! What does this have to do with the dark satellites?"

"What happens when something is dark?" Red Crab asks.

······一种用来寻找光······

... one looking for light ...

"那还用问？亮的时候它反射光和热，暗的时候它吸收光和热。"

"所以，寻找光线的望远镜会被强光蒙蔽，而寻找热量的望远镜会被深色的人造卫星干扰。那么，我们会怎样呢？我们会迷路吗？"

"Of course, when bright it reflects light and heat, and when dark it absorbs heat and light."

"So, telescopes that look for light are blinded by a bright light, and telescopes that look for heat will be confused by dark satellites. And what about us, losing our way?"

"我们要求人类尊重我们流传了数百万年的导航方式。"

"是的，请从人类那里了解一下，我们这些利用夜空导航的生物，现在如何知道什么时候、到哪里去生育宝宝！"

……这仅仅是开始！……

"We all need to demand respect for our way of navigating, one that is millions of years old."

"Yes, please find out from people how we, the creatures that use the night sky to navigate, are now going to know when and where to go, to have our little ones!"

... AND IT HAS ONLY JUST BEGUN!...

······ 这仅仅是开始！ ······

... AND IT HAS ONLY JUST BEGUN! ...

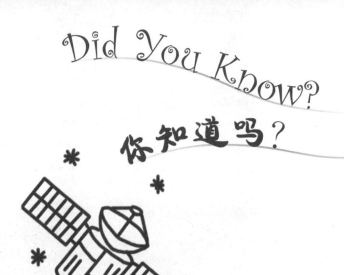

In 2020, an estimated 20,000 satellites are orbiting the Earth. One company filed for launching 30,000 extra satellites. Observatories in Chile are confused by 600 to 700 space objects observed at any given time.

2020 年, 估计有 2 万颗人造卫星绕地球运行。一家公司申请发射另外 3 万颗人造卫星。智利观测站在任何给定时间都会被观测到的 600 到 700 个太空物体干扰。

SpaceX calculated it needs a fleet of 1,584 satellites in orbit to provide near-continuous internet service to the world's populated areas. As these satellites fly close to the Earth they block out the night skies.

据 SpaceX 计算, 该公司需要 1584 颗在轨卫星, 才能向世界人口稠密地区提供近乎不间断的互联网服务。因为这些人造卫星近地飞行, 它们会遮挡夜空。

The blackpoll warbler takes a ride on trade winds, sailing from the north-eastern USA to South America in a hundred hours – completely guided by the stars. On the way back, it stops on land to rest and refuel.

黑顶白颊林莺搭着信风，从美国东北部飞行到南美洲需要 100 个小时，飞行过程中完全依靠星星引路。在返回途中，它会在陆地上停下来休息并补充能量。

The Mexican free-tailed bat travels 70 kilometres each night to feed on moths and mosquitos. It uses the stars, landmarks, and the smell of their fellow cave bats, to find its way home.

墨西哥无尾蝙蝠每晚要飞行 70 千米以捕食飞蛾和蚊子。它利用星星、地标和同一洞穴的同伴的气味找到回家的路。

为了觅食，沙漠蚁会爬到距中心巢穴 500 米以内的地方，通过计算步数来精确地记住它们走了多远。它们还用太阳的偏振光导航。

Desert ants travel up to 500 m away from their central nest sites to search for food, and remember exactly how far they've gone by counting their steps. They also navigate via polarised light patterns from the Sun.

科学研究为光污染对野生动植物的影响提供了新见解。目前已经证实，蜣螂的眼睛和大脑已经进化到能够对远在太阳系之外的视觉信号作出反应。

Research offers new insights into the effects of light pollution on wildlife. It has been established that the eyes and brains of dung beetles have evolved to respond to visual cues from far beyond our solar system.

Migrating birds do not pay attention to individual stars but rather to the rotation of star patterns. This will, for instance, enable birds to determine where north is, and they then use this information to fly south.

候鸟不是观察单个星星，而是观察群星图案的转动。例如，鸟类能够据此确定哪里是北方，然后利用这些信息向南飞行。

Seals use specific lodestars as navigational cues to venture far from shore, in instances where they lack terrestrial landmarks. The animals are able to identify a single star in the Northern Hemisphere night sky.

在缺少陆地地标的情况下，海豹会用北极星这颗特定的恒星作为导航信标，到远离海岸的地方探险。这种动物能够识别北半球夜空中的单颗恒星。

Would you like to be able to read the stars in the sky to find your way?

你愿意通过观察天上的星星来寻路吗？

What advice do you have for the owners of satellites?

你对人造卫星的拥有者有什么建议？

Do the creatures that have had guidance from the sky for millions of years have the right to this guidance forever, or can it be taken away?

几百万年来利用天空导航的生物是否有权利永远用这种方式导航，或者这种权利可以被剥夺吗？

Could you have imagined that crabs and beetles, with their small brains, are able to process so much information?

你能想象螃蟹和甲虫的大脑虽小，却能处理如此多的信息吗？

How many animals do you know that are guided by the Sun, the Moon, the stars, and the Milky Way to travel around? Undertake some research and look for at least 20 such animals. You will notice that not every animal does this, but that there are nevertheless many that have the ability to do so. Share this list with friends and family members and find out what everyone thinks about the confusion satellites cause amongst these animals guided by the sky.

你知道有多少动物是在太阳、月亮、星星和银河系的指引下四处旅行的？做些研究，至少找出 20 种这样的动物。你会注意到虽然不是所有的动物都这么做，但还是有很多动物有这样的能力。把这份名单分享给朋友和家人，看看他们每个人都怎么看待人造卫星给靠天空导航的动物造成的混淆。

学科知识
Academic Knowledge

生物学	红蟹用鳃呼吸；蟹从最初的幼体成长为一种像虾一样的动物，叫作大眼幼体；红蟹吃树叶、花和种子；5 000种不同种类的蜣螂专门以粪便为食；蜣螂的复眼。
化 学	粪便的气味取决于64种不同的化学成分。
物 理	红蟹避开阳光，躲进洞里，保持潮湿，封住洞口长达3个月；蜣螂一个晚上能掩埋相当于自己体重250倍的东西，也可以滚动10倍于自己体重的东西；蜣螂有很好的嗅觉；光和热的吸收。
工程学	蜣螂可作为研究气候变化，特别是干旱、火灾和人类活动影响的生物指标；蜣螂能改善卫生和健康，能使灌木蝇减少90%；在宇宙中寻找光或热；望远镜。
经济学	通信业是一个高增长行业，即使在危机时期也是如此。
伦理学	在有其他选择的情况下，我们能仅仅为了金钱和通信的目的而遮挡天空吗？
历 史	蜣螂在古埃及的地位；伊索寓言《鹰与甲虫》，安徒生童话《屎壳郎》；弗兰兹·卡夫卡的《变形记》中的主角变成了甲虫。
地 理	地球同步轨道（与地球自转同步）；红蟹生活在森林里；红蟹的迁徙与月亮的圆缺有关；雌蟹会在大潮，也就是下弦月的时候，把卵产到海里；位于印度尼西亚和澳大利亚之间的圣诞岛；夜间活动的非洲蜣螂以银河系作为导航工具，有的还根据月光的偏振类型导航。
数 学	计算和比较效率（每晚食用自身质量250倍的食物；拉动自身质量10到1 000倍的物体）。
生活方式	中国蜣螂是一味中药；随心所欲，不考虑别人，把自己的观点强加于人。
社会学	蜣螂偷其他蜣螂的东西；圣甲虫传递转变、更新和复活的思想；用绿色石材制作圣甲虫，将其放在死者的胸膛上；在西部荒野，只有自己的规则才适用。
心理学	对他人不顾影响的行为的感受；自我中心主义，只关注自己。
系统论	由于引进了黄疯蚁，圣诞岛上的红蟹大量死亡；在当地特有的麦克里尔鼠灭绝后，红蟹主导着这个岛屿；建设红蟹桥，以确保它们安全过马路；蜣螂对环境高度敏感；蜣螂在养分循环、土壤通气和害虫、寄生虫的生物防治中发挥着重要作用；蜣螂清除粪便，否则那里就会成为苍蝇的滋生地。

情感智慧
Emotional Intelligence

蜣螂

蜣螂花时间研究红蟹的习性，并艰难跋涉，亲自拜会红蟹。尽管他从未见过红蟹，但他与红蟹进行了面对面的交谈，并有信心讨论红蟹的后代。他称赞红蟹的独特能力。他谦逊而又恰如其分地评价自己的能力。他简练地分享信息，表述精确，他的解释和结论也比较简单。然而，当红蟹不明白的时候，蜣螂需要进一步解释。在对话的过程中，蜣螂得出结论：人们应该尊重他们古老的利用夜空的方式。

红蟹

红蟹分享令人惊叹的数字，是想给人留下深刻印象。她很有自知之明，说话的方式使人宽慰。她知道自己的能力非凡，并不遮掩。她很警觉，注意到蜣螂分享的事实，并坦然地接受他的热情。虽然以前从未见过面，但他们惺惺相惜。既然有相同的兴趣，红蟹就问蜣螂关于"新星"的事。她不愿意接受事实，并质疑蜣螂提供的信息。她主要关心所有受影响生物的利益，并寻求对此问题的更好的共识。

艺术
The Arts

天空中的人造卫星就像一串串珍珠。让我们用电脑来展示夜空。找一个延时摄影程序来展示星星的运动。接下来，在美丽的夜空上添加50 000颗人造卫星。用计算机程序不断地增加人造卫星的数量，直到模拟出50 000颗人造卫星。这些人造卫星严重阻挡视线，"破坏"夜空，使那些靠太阳、星星、月亮和银河系导航并以此为生的生物混淆信号。

思维拓展
Systems: Making the Connections

从地球上观察到的宇宙景象，几千年来一直是航海家和探索者的向导。它让夏威夷水手能够穿越太平洋；它还是很多原住民的向导，如非洲的多贡人，根据它来确定种植和收获的时间；它使阿拉伯和中国的天文学家能够超越地球，直探太空，发现地球是圆的。望远镜有两种：一种是在黑暗中寻找光，另一种是在寒冷的太空中寻找热。这两种类型的望远镜不仅用于研究银河系、黑洞或者宇宙形成理论，还用于发现新的未知天体信号。它们的主要功能之一是及时识别可能对地球生命构成威胁的陨石。这一至关重要的警戒职能，以及所有其他需要完成的科学任务，都日益受到人造卫星的阻碍。它们有的反射光线，像天空中的珍珠一样闪烁；有的则被涂暗并吸收热量，影响精确观测靠近的陨石。人们不禁要问，发展通信业的同时，怎么能忽视人造卫星的副作用？当第一颗人造卫星发射升空时，所有人都在关注它的积极影响。随着军备竞赛的展开，发展人造卫星是一件关乎自豪感和国防的事情，因此没有人提出任何问题。如今，人造卫星已经超越了民族自豪感的范畴。人造卫星已经变得具有战略意义，其所有权也从公有转为私有。只有如今数万颗人造卫星计划铺开，世界才开始意识到人造卫星的严重副作用。现在，通过计算机编程和人工智能，成千上万张天空图片可以粘贴在一起，这让我们清楚地看到问题的严重性。四处漂浮的废弃太空材料碎片正是一个重大挑战。人类将不得不接受这样一个事实：数十万颗人造卫星将很快被部署到各个空域。这不仅会影响我们探索宇宙，还会影响成千上万种动物的导航——这些动物可能再也找不到回它们的栖息地或繁殖地的路。

动手能力
Capacity to Implement

如何区分恒星和人造卫星？人造卫星的运行轨迹是直线，在夜空中穿过需要几分钟。流星划过夜空只要几分之一秒。此外，人造卫星会有规律地变亮和变暗。这就是我们如何用肉眼区分它们的方法。现在，既然你能识别人造卫星，你能数一下有多少颗人造卫星吗？

故事灵感来自

This Fable Is Inspired by

凯特琳·M·凯西
Caitlin M. Casey

2007 年，凯特琳·M·凯西在美国亚利桑那大学获得物理学、天文学和数学专业的理学学士学位。2010 年，她在英国剑桥大学圣约翰学院攻读天文学博士学位，研究早期宇宙中的超亮红外星系。她在夏威夷大学天文学研究所和加利福尼亚大学尔湾分校获得博士后研究职位，并被任命为得克萨斯大学奥斯汀分校天文系助理教授。她致力于研究星尘星系、红外线和亚毫米波观测以及恒星形成。

图书在版编目(CIP)数据

冈特生态童书.第七辑:全36册:汉英对照 /
(比)冈特·鲍利著;(哥伦)凯瑟琳娜·巴赫绘;
何家振等译.—上海:上海远东出版社,2020
ISBN 978-7-5476-1671-0

Ⅰ.①冈…　Ⅱ.①冈…②凯…③何…　Ⅲ.①生态
环境–环境保护–儿童读物—汉英　Ⅳ.①X171.1-49

中国版本图书馆CIP数据核字(2020)第236911号

策　　划 张蓉
责任编辑 祁东城
封面设计 魏　来　李　廉

冈特生态童书
天上的珍珠
[比]冈特·鲍利　著
[哥伦]凯瑟琳娜·巴赫　绘
何家振　译

记得要和身边的小朋友分享环保知识哦!
八喜冰淇淋祝你成为环保小使者!